D1105160

West Georgia Regional Library System
Regional Office

DISCARD

THE
BRAIN

CHELSEA HOUSE
PUBLISHERS
A Haights Cross Communications Company ®
Philadelphia

First hardcover library edition published
in the United States of America in
2006 by Chelsea House Publishers,
a subsidiary of Haights Cross Communications.
All rights reserved.

A Haights Cross Communications Company ®

www.chelseahouse.com

Library of Congress Cataloging-in-Publication
applied for.
ISBN 0-7910-9014-0

Project and realization
Parramón, Inc.

Texts
Adolfo Cassan

Translator
Patrick Clark

Graphic Design and Typesetting
Toni Inglés Studio

Illustrations
Marcel Socías Studio

First edition - September 2004

Printed in Spain
© Parramón Ediciones, S.A. – 2005
Ronda de Sant Pere, 5, 4ª planta
08010 Barcelona (España)
Norma Editorial Group

www.parramon.com

The whole or partial reproduction of this work by
any means or procedure, including printing,
photocopying, microfilm, digitalization, or any
other system, without the express written
permission of the publisher, is prohibited.

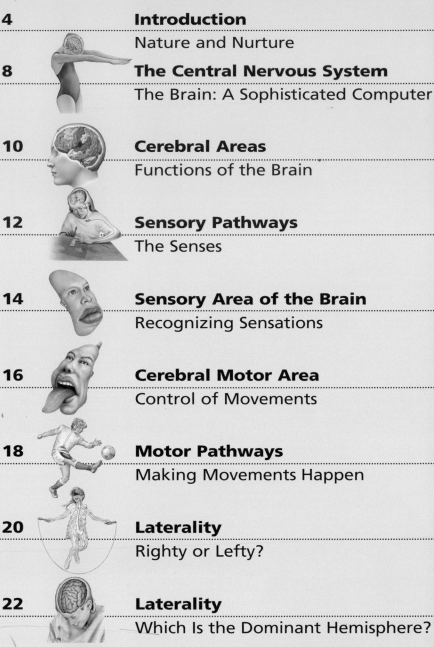

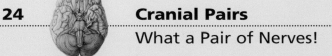

EVERYTHING UNDER CONTROL

This book explores the brain and its higher functions. It aims to provide readers with some basic information about the amazing abilities of our nervous system and our brain in particular. The brain allows us to understand what goes on around us, to reason, to experience emotions, and to remember. If we understand the way the nervous system is organized and how it works, we will be better able to keep our brains as healthy as possible.

Our goal was to create a book that is practical, educational, challenging, and at the same time, entertaining for the reader. We hope that our readers will consider this mission accomplished.

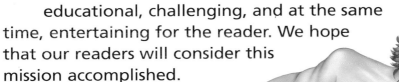

NATURE AND NURTURE

The nervous system is formed by a specialized cell, called a neuron, that is able to recognize different kinds of stimuli and send messages about them to other cells to achieve certain results, such as the contraction of a muscle.

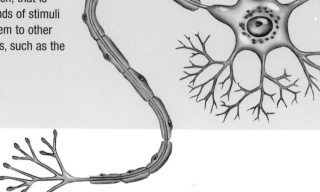

THE HIGHER FUNCTIONS

Our body carries out many functions that are absolutely necessary for life. Among these important functions are digestion, respiration, and blood circulation. These functions are common to nearly all members of the animal kingdom. But there are other, more complex functions that occur in human beings such as learning, language, emotion, memory, and abstract thought.

It is worth noting that, unlike plants, which only respond to a certain number of physical and chemical stimuli in the environment, animals, including humans, respond to many different agents. This is because animals have a specific organ system that can recognize many outside stimuli and generate responses to help the animal adapt to changes in the environment. This organ system is known as the nervous system.

THE NERVOUS SYSTEM

Even in primitive animals, some special cells can respond to external stimuli. Over the course of evolution, these cells have developed into nerve cells, or neurons, that can recognize stimuli. Once a neuron recognizes stimuli, it generates signals and sends messages to other cells. The recipient cells will then carry out a specific function, such as making a muscle contract or having a gland secrete a hormone.

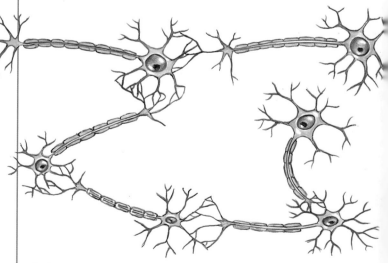

Each neuron is connected to other neurons, sometimes in groups of hundreds, forming the very complicated network that makes up the nervous system.

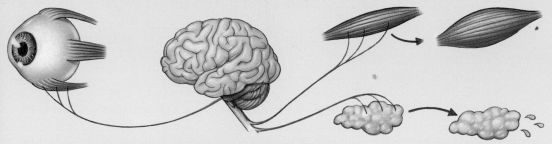

The nervous system's basic functions include registering external stimuli through sense organs; generating responses, such as muscle contractions or glandular secretions; and adapting in the best possible way to changes in the environment.

Less developed animals have a much simpler nervous system than ours. It is made up mainly of a series of nerve cells joined by cords that connect to a small, primitive brain.

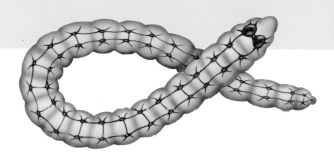

As animals get more complex, neurons increasingly interact to create different types of nervous systems. The reticular nervous system found in simple animals, for example, is a nerve network inside the body. Information flows through it whenever any part of the body senses a stimulus. The nervous system found in animals with long, segmented bodies like earthworms and arthropods has masses of neuron bodies arranged in pairs in each segment. These neuron bodies interact by means of nerve fibers that form a primitive brain.

The most complex type of nervous system is found in vertebrates (animals with backbones), including humans. The nervous systems of humans and other vertebrates consist of a central nervous system, which is controlled by the brain and the spinal cord. There is also a peripheral nervous system, which is made up of cranial pairs and many spinal nerves. The peripheral nerves send signals about sense stimuli to the brain, or send the orders generated in the brain to the muscles that carry them out.

THE CEREBRUM: THE CENTER OF THE MIND

In the brain, there are various groups of neurons that make up nerve centers. Some are responsible for controlling vital functions such as breathing or heartbeat. These centers are found in all animals, from the most primitive creatures to complex humans. In complex animals, the brain is more complicated and has structures that play a role in behavior. For example, "instinct" is an animal's way of reacting to different external stimuli through reflex action. As we go higher up the scale of evolution, the size of the brain also increases, especially the size of the most important component, the cerebrum.

The human nervous system stands out from that of other animals because of the size of the brain in relation to body mass.

The brain is made up of two large symmetrical masses called hemispheres, which are linked by bundles of nerve fibers that allow the hemispheres to communicate. Each hemisphere has a central cavity filled with fluid, called the ventricle. An inner layer, just outside the ventricle, consists of a cluster of neurons that create the centers that handle basic functions. An intermediate layer, called white matter, is made of bundles of associative nerve fibers. The outer surface layer, called gray matter, is made up of many interconnected neurons.

For its size and sophistication, the human brain has no equal in the animal kingdom. The cerebral cortex is so extensive that, given the limited space within the cranium, it has to make many curves and folds, which give the surface of the brain a very wrinkled appearance. Learning, language, emotion, memory, and abstract thought occur in the cerebral cortex.

FEELING AND THINKING

Information that the sense organs receive from the outside world and inside the body is sent to the cerebral cortex. Part of this information is processed automatically, without thinking about it. However, information is also processed by our conscious perceptions based on what we see, hear, and feel. The orders that the corresponding nerves carry to the muscles, which become voluntary movements and actions, are generated in the cerebral cortex.

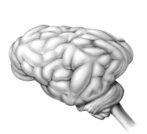
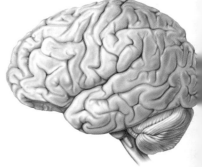

The differences between the brain of an animal such as a dog and that of a human being are based not only on size, but also on another quality: The canine brain is much smoother, while the much larger human cerebral cortex has many folds.

Although our mental faculties have a biological basis, environmental factors also play a role in development. During childhood and until the end of adolescence, connections between neurons form, increasing intellectual capacity.

Besides the sensory and motor functions, the nervous system performs much more sophisticated tasks. It controls learning, memory, emotions, motivations, perception, language, planning, imagination, and abstract thought. How do these complex functions arise? The answer is that we still don't know exactly how these things occur, although scientists are learning more details about how the brain works every day.

NATURE VERSUS NURTURE

How much of the development of complex functions depends on biological factors, and how much depends on external factors such as the environment? Although both factors are important, it is a difficult question to answer. Neurons are not distributed by chance. They are organized in a very precise way, so that they form complicated and interconnected pathways, or circuits, that can handle a variety of functions. This order is already determined from the moment the egg is fertilized by the sperm, so there is undoubtedly an innate basis for it. But today we know that external factors—especially during early childhood development—influence how

neuronal circuits are formed. What happens around us, the way our parents raise us, what we learn in school, and all the stimuli present in the world in which we live work together to shape our brains. Thus, our abilities depend, in large part, both on the genes we inherit from our parents, many of which are common to the entire human species, and to our everyday experiences.

The human nervous system is very immature when a baby is born. It needs appropriate stimulation to develop properly. Children's games play a big role in this development.

THE BRAIN:
A SOPHISTICATED COMPUTER

The brain is part of the central nervous system. Sophisticated structures in the brain control all the higher functions and are responsible for our voluntary conscious actions as well as the automatic functions of our bodies. The brain controls our relationships with the world, and the complex intellectual processes that allow us to be conscious of ourselves and of everything around us.

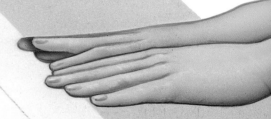

The cerebral cortex, center of intelligence

Intelligence is often defined as the ability to adapt to new situations, something that is key for survival. This is not a fixed ability. Instead, it is a set of intellectual faculties that can be carried out because of the complex connections between the different parts of the cerebral cortex, the layer of nerve cells that covers the human brain.

THE COMPLEXITY OF THE BRAIN

The brain, like the rest of the structures that make up the nervous system, is composed of two types of cells: neurons and glia. Neurons recognize stimuli and transmit messages throughout the body. Glia, or support cells, help provide nutrition and protect the neurons. The brain contains around a billion neurons, all perfectly interconnected so they can work in a coordinated way with great precision.

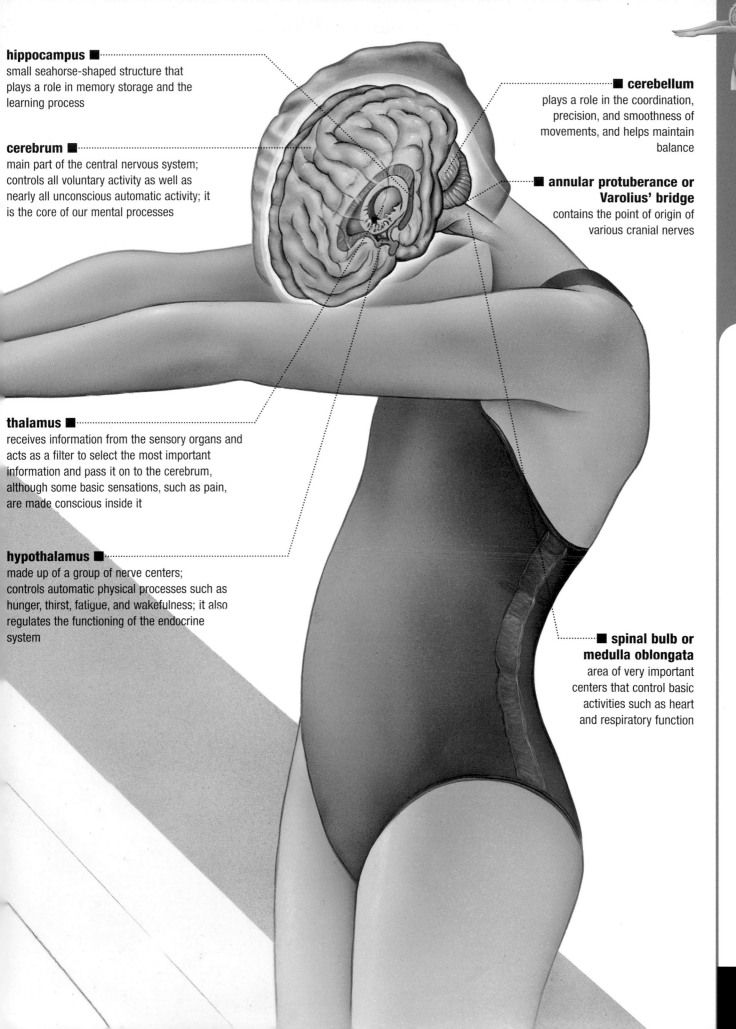

hippocampus ■
small seahorse-shaped structure that plays a role in memory storage and the learning process

cerebrum ■
main part of the central nervous system; controls all voluntary activity as well as nearly all unconscious automatic activity; it is the core of our mental processes

thalamus ■
receives information from the sensory organs and acts as a filter to select the most important information and pass it on to the cerebrum, although some basic sensations, such as pain, are made conscious inside it

hypothalamus ■
made up of a group of nerve centers; controls automatic physical processes such as hunger, thirst, fatigue, and wakefulness; it also regulates the functioning of the endocrine system

■ **cerebellum**
plays a role in the coordination, precision, and smoothness of movements, and helps maintain balance

■ **annular protuberance or Varolius' bridge**
contains the point of origin of various cranial nerves

■ **spinal bulb or medulla oblongata**
area of very important centers that control basic activities such as heart and respiratory function

FUNCTIONS OF THE BRAIN

The brain is the most important organ of the central nervous system, and is responsible for all of the higher functions. Although its operation is very complex and still not entirely understood, scientists have been able to identify several areas of the brain that control basic activities, such as movement, language, and vision. In the future, it may be possible to create a "map" that indicates which parts of the brain control the major mental functions.

premotor area ■
mainly controls head and eye movements

frontal lobe ■
main area where various mental functions are developed, and the center of behavior

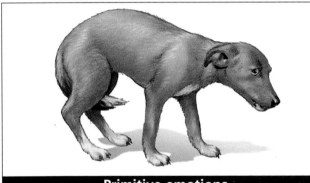

Primitive emotions

The brain areas related to emotions are the most primitive, from an evolutionary point of view. Faced with certain stimuli, such as those that cause fear, we react in a way that is similar to how less developed animals react.

motor area ■
controls all voluntary movements of the body, from head to toe

CEREBRAL CONNECTIONS

Although parts of the cerebral cortex handle specific functions, the truth is that the activity of the brain is very complex. Many of the higher functions, such as intelligence and judgment, do not have a specific location. Instead, they depend on interactions between different parts of the brain. It should be stressed that the development of intellectual activities, in which the brain as a whole takes part, depends both on genetic inheritance and on many environmental factors, including education.

sensation area ■
registers and interprets sensory information from all over the body, both inside and on the surface

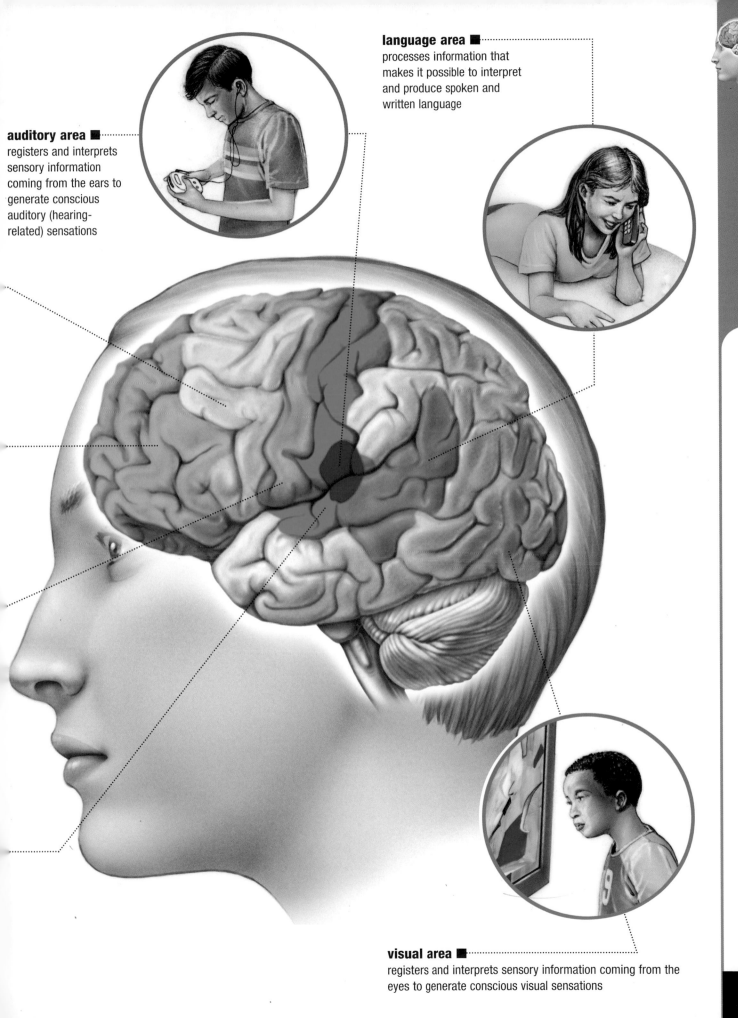

auditory area ■
registers and interprets sensory information coming from the ears to generate conscious auditory (hearing-related) sensations

language area ■
processes information that makes it possible to interpret and produce spoken and written language

visual area ■
registers and interprets sensory information coming from the eyes to generate conscious visual sensations

THE SENSES

Sensory stimuli corresponding to touch, pressure, pain, and temperature are registered on the surface of the body or inside, then must travel a long path to be perceived. The sensory receptors that detect stimuli generate nerve impulses that are sent through sensory nerve fibers to the spinal medulla, and along the length of specific pathways to the cerebral cortex, where sensations are made conscious.

cerebral cortex ■
sensory stimuli are decoded and turned into conscious bodily sensations

thalamus ■
some stimuli are filtered, and the rest are sent to nerve fibers that are directed to the cerebral cortex

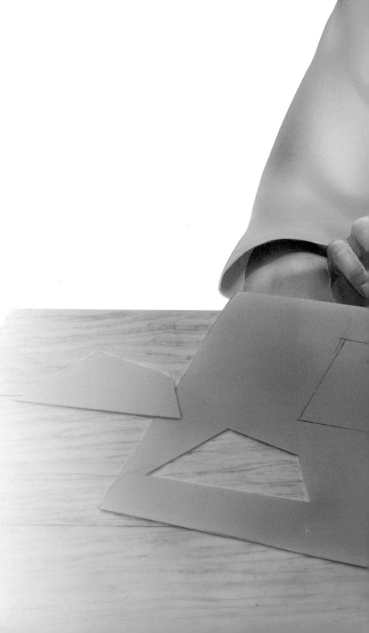

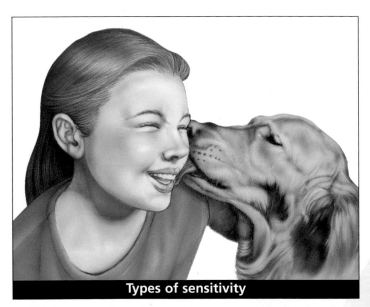

Types of sensitivity

Not all sensory stimuli are perceived equally, since they may differ in both nature and function, and they may even be recognized by different areas of the brain. There are two main categories of sensitivity: fine and coarse. Fine sensitivity, also called epicritical, is more precise. It is made conscious in the cerebral cortex, and allows us to quickly detect tactile (touch) stimuli. Coarse sensibility, also known as protopathic, is essential for its alarm function. It is less refined and not very localized. It is made conscious in the thalamus, and mainly recognizes pain and thermal (temperature-related) stimuli.

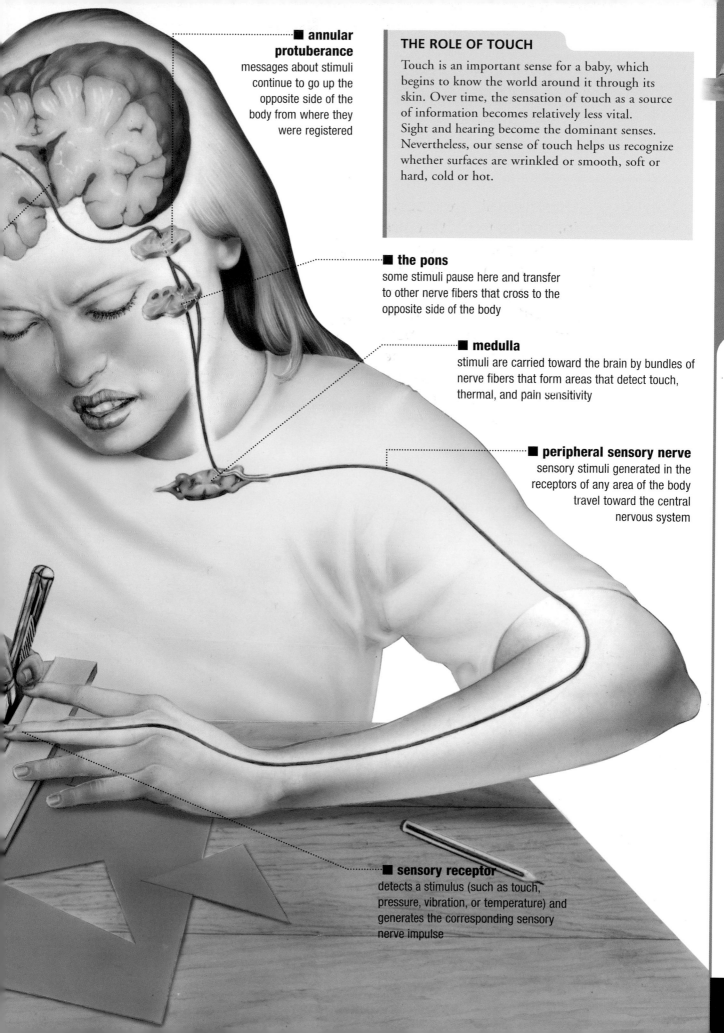

annular protuberance
messages about stimuli continue to go up the opposite side of the body from where they were registered

THE ROLE OF TOUCH
Touch is an important sense for a baby, which begins to know the world around it through its skin. Over time, the sensation of touch as a source of information becomes relatively less vital. Sight and hearing become the dominant senses. Nevertheless, our sense of touch helps us recognize whether surfaces are wrinkled or smooth, soft or hard, cold or hot.

the pons
some stimuli pause here and transfer to other nerve fibers that cross to the opposite side of the body

medulla
stimuli are carried toward the brain by bundles of nerve fibers that form areas that detect touch, thermal, and pain sensitivity

peripheral sensory nerve
sensory stimuli generated in the receptors of any area of the body travel toward the central nervous system

sensory receptor
detects a stimulus (such as touch, pressure, vibration, or temperature) and generates the corresponding sensory nerve impulse

RECOGNIZING SENSATIONS

Impulses coming from sensory receptors all over the body arrive at a specific area of the cerebral cortex. There, impulses are processed and made conscious. Since each sense receptor sends its messages to a specific point, we get an odd view of the body by looking at the sensory areas of the cerebrum. Larger sensory areas are dedicated to the most sensitive parts of the body.

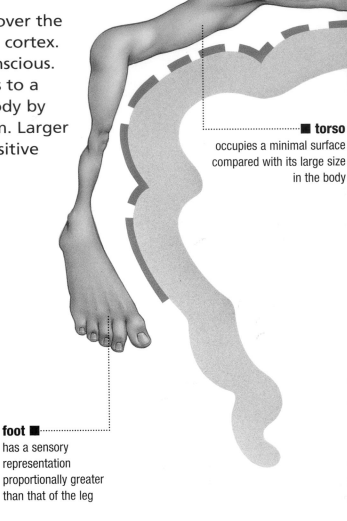

■ torso
occupies a minimal surface compared with its large size in the body

foot ■
has a sensory representation proportionally greater than that of the leg

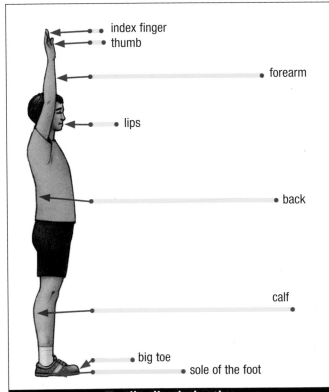

index finger
thumb
forearm
lips
back
calf
big toe
sole of the foot

Tactile discrimination

Tactile (touch) sensitivity of different parts of the body depends on how many receptors are present on the surface of the skin. There are so many receptors in the fingertips that it is possible to tell the difference between two stimuli separated by fractions of an inch, while in some parts of the back, two stimuli that are separated by an inch or more will be felt as one. This illustration shows the minimum distance needed to differentiate two points of the body that are touched separately.

WHAT IS A HOMUNCULUS?

In neuroanatomy, a homunculus is a way of showing how much tactile sensation each part of the body takes up in the cerebral cortex. The hands and face, for example, look huge, because they have so many sensory receptors and can detect even fine touches. As you can see, the proportion of an area depicted in a homunculus often has no relation to the actual size of a particular part of the body.

SENSORY RECEPTORS

It is estimated that there are around 4 million receptors on the surface of the body for the sensation of pain, 500,000 for pressure, 150,000 for cold, and 16,000 for heat.

sensory area of the brain ■ ┈┈┈┈┈┈┈┈┈┈┈┈┈┈┈┈┈┈┈

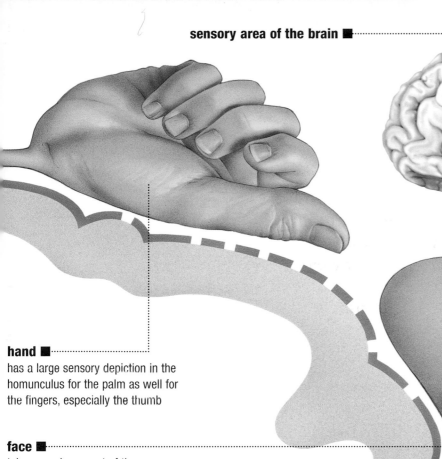

hand ■┈┈┈┈┈┈┈┈┈
has a large sensory depiction in the
homunculus for the palm as well for
the fingers, especially the thumb

face ■
takes up a large part of the sensory
area, since the whole surface of the
face is very sensitive

lips ■
proportionally, this is the most sensitive
part of the body

teeth and gums ■┈┈┈┈┈┈┈┈┈
these are very sensitive areas,
especially in response to pain

tongue ■┈┈┈┈┈┈┈┈┈
has great sensitivity to touch,
pressure, and temperature, which is
unrelated to its ability to detect taste
sensations

pharynx ■┈┈┈┈┈┈┈┈┈
the throat is a very sensitive
body area

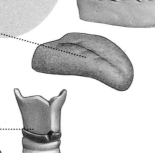

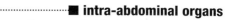

■ **intra-abdominal organs**
the organs inside the abdomen are more
sensitive than the inner organs of other
parts of the body

CONTROL OF MOVEMENTS

All of the voluntary movements that we make begin in a specific area of the cerebral cortex, where nerve cells generate the appropriate orders to the muscles that carry out actions. Since there is an exact correlation between each sector of this area and motion control, motor ability can be represented by a distorted face, as seen here.

face ■
has a large presence in the motor area, evident in the great mobility of the muscles that control the wide variety of facial expressions

lips ■
have a very significant area, since they are very mobile and play a very active role in eating and talking

tongue ■
corresponds to a disproportionately large part of the motor area, due to its important role in the process of chewing and (especially) in spoken language

PROGRAMMING OF MOVEMENTS

If the motor area of the cerebral cortex is indeed responsible for generating body movements, then it is connected with other areas of the brain that are in charge of planning such movements. When we move, we do it with intention, in response to certain stimuli. Movements are simply a way to achieve some purpose, such as grabbing an object. Planning movements is very complex. The motor area is only the final station at which an intention is turned into movement.

torso ■
has a reduced presence compared to its size in the body, since it tends to move as a unit

■ **hand**
occupies a large part of the motor area because of the great mobility of the palm and, especially, the precision of finger movements

foot ■
has a motor area presence proportionally much larger than that of the leg

GIANT CELLS

In the motor area, some neurons are much bigger than the rest. These are known as "giant pyramidal cells." They were discovered in 1879 by Russian anatomist Vladimir A. Betz. They are responsible for generating the nerve signals that give rise to our voluntary movements.

■ **larynx**
is the organ of voice and speech and, given the importance of the vocal cords in producing the sounds of spoken language, it takes up a good deal of space in the motor area

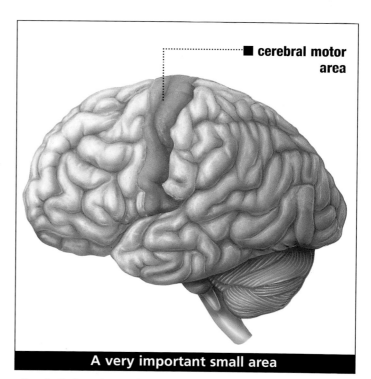

■ **cerebral motor area**

A very important small area

Despite its importance, the motor area represents a small part of the cerebral surface: It is barely the size of a finger.

MAKING MOVEMENTS HAPPEN

Impulses generated in the neurons of the cerebral motor area, intended to produce the wide variety of movements we make voluntarily, must travel a long path to reach their targets. They travel from the brain to the brain stem. There, many of the motor neuron extensions cross over to the opposite side of the body, move through the spinal cord, and finally, pass to the peripheral nerves that reach the muscles that carry out the orders of the brain.

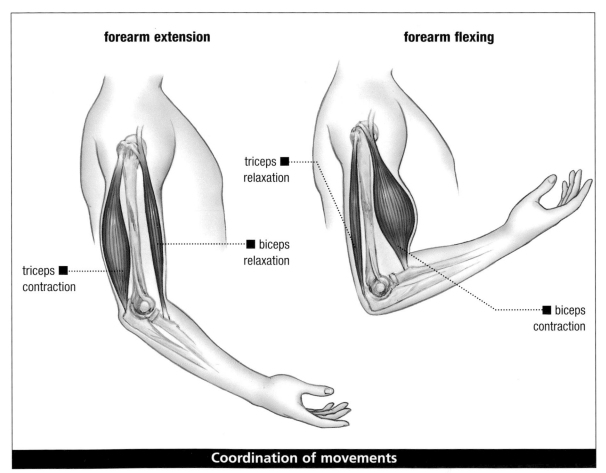

forearm extension

forearm flexing

triceps ■
relaxation

■ biceps
relaxation

triceps ■
contraction

■ biceps
contraction

Coordination of movements

To carry out a movement is a more complex process than we might imagine. For example, extending or flexing the forearm requires coordination of the biceps and triceps muscles: One of these must contract while the other relaxes. To carry out more complex movements, such as walking, jumping, or running, several muscles need to contract or relax in a specific way. Fortunately, it is usually enough for us to intend to make a certain movement for our brains to send the instructions to the corresponding muscles.

■ **cerebral cortex**
motor stimuli are generated in specialized neurons that are known, because of their appearance, as pyramidal cells

■ **pyramidal pathway**
the continuations of pyramidal cells descend down the brain stem and make up a motor pathway

■ **annular protuberance**
stimuli continue to travel down the same side of the cerebral hemisphere where they were generated

■ **the pons**
80% of the nerve fibers cross over to the other side of the body, while the rest follow their path on the same side

■ **direct pyramidal bundle**
the fibers that do not cross to the other side in the spinal bulb form a cord that goes down through the front part of the spinal medulla

■ **crossed pyramidal bundle**
fibers that have crossed to the other side in the spinal bulb form a cord that goes down through the side of the spinal medulla

■ **spinal cord**
nerve fibers in the spinal cord send orders to the neurons that make up the motor nerves

■ **peripheral motor nerves**
impulses travel to the muscles that need to contract to carry out movements

FINE MOVEMENTS

Although the cerebrum is mainly responsible for our voluntary movements, there are other brain structures that also take part. The cerebellum stands out as one of these. The cerebellum plays a role in the coordination of muscular contractions and regulates movements, making sure they are carried out as smoothly and precisely as possible.

RIGHTY OR LEFTY?

As nerve pathways cross each other on their route between the brain and the rest of the body, each cerebral hemisphere controls the movements and sensitivities of the opposite half of the body. One cerebral hemisphere is often "dominant" with respect to the other: Usually, the left cerebral hemisphere is dominant, resulting in greater motor control on the right side of the body (right-handed people). The right hemisphere is dominant in some people, who have greater motor control on the left side (left-handed people).

PERCENTAGES

Traditionally, scientists believed that around 90% of people are right-handed and 10% left-handed, but some studies show that between 10% and 15% of people have a mixed laterality. This means that they have a dominant right hand and a dominant left foot (or vice versa). About 5% of individuals are ambidextrous—having the same ability in motor control on both sides of the body.

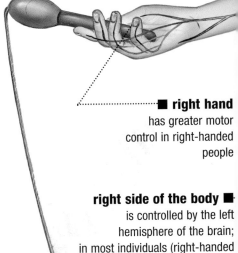

right hand
has greater motor control in right-handed people

right side of the body
is controlled by the left hemisphere of the brain; in most individuals (right-handed people), has greater precision in movements

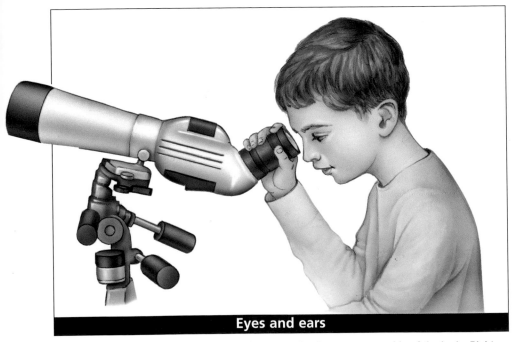

Eyes and ears

Laterality shows up in other ways besides motor dominance on one side of the body: Right-handed people tend to prefer to use their right eyes for close observation or to look through an optical apparatus, and they tend to use their right ears to listen to a sound with maximum attention, while left-handed people do the opposite.

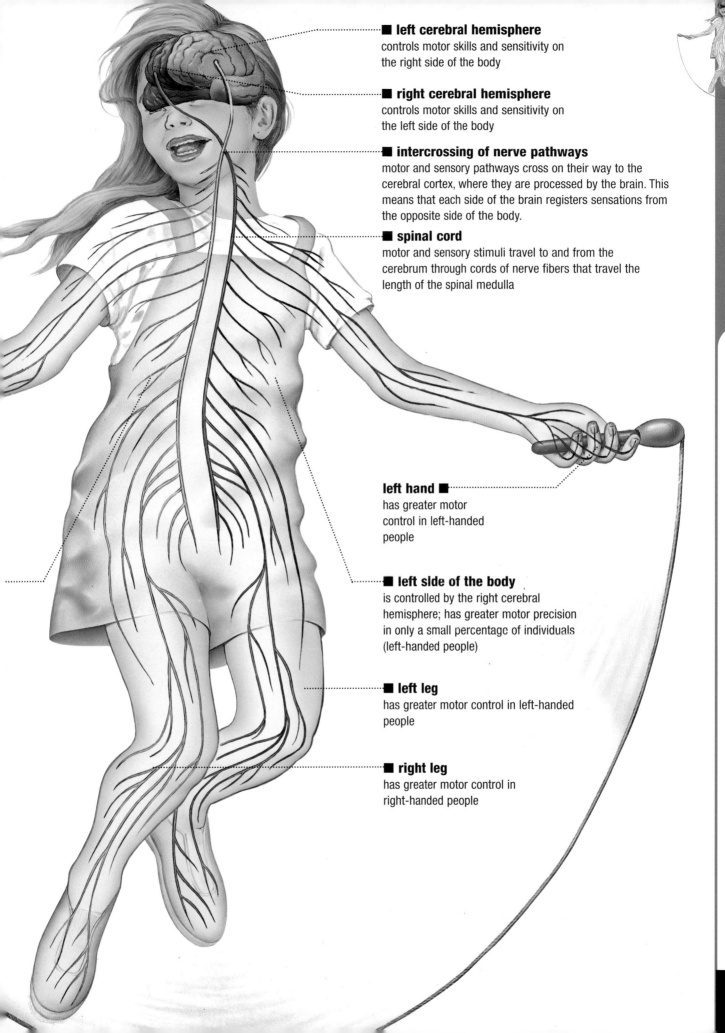

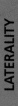

■ **left cerebral hemisphere**
controls motor skills and sensitivity on
the right side of the body

■ **right cerebral hemisphere**
controls motor skills and sensitivity on
the left side of the body

■ **intercrossing of nerve pathways**
motor and sensory pathways cross on their way to the
cerebral cortex, where they are processed by the brain. This
means that each side of the brain registers sensations from
the opposite side of the body.

■ **spinal cord**
motor and sensory stimuli travel to and from the
cerebrum through cords of nerve fibers that travel the
length of the spinal medulla

left hand ■
has greater motor
control in left-handed
people

■ **left side of the body**
is controlled by the right cerebral
hemisphere; has greater motor precision
in only a small percentage of individuals
(left-handed people)

■ **left leg**
has greater motor control in left-handed
people

■ **right leg**
has greater motor control in
right-handed people

WHICH IS THE DOMINANT HEMISPHERE?

Besides registering sensitivity and controlling the movements of the opposite side of the body, each of the two cerebral hemispheres handles several functions and intellectual capacities. In general terms, the dominant hemisphere (corresponding to the left hemisphere in right-handed people and to the right hemisphere in left-handed people) is more logical and concerned with language, while the other hemisphere is more emotional and artistic.

music ■
musical sense, sensitivity to music, ability to compose, and ability to play musical instruments

emotions ■
emotional content, facial and physical expression of emotions, ability to control and dominate emotions

Left-handed geniuses

Left-handed people are a small part of the population, but it is noteworthy that in some activities, especially those related to music and general creativity, the number of left-handed individuals is proportionally higher. Over the course of history, there have been many outstanding left-handed creative and scientific geniuses, including Leonardo da Vinci (shown here), Benjamin Franklin, and Charlie Chaplin.

art ■
artistic interest and/or talent (such as drawing and painting, literature, and sculpture), creative ability, imagination and fantasy

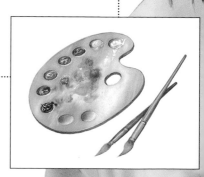

spatial ability ■
sense of direction in space, calculation of direction and distance, ability to manipulate relative positions of objects in space, depth perception

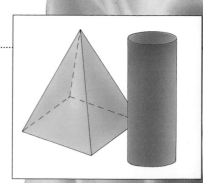

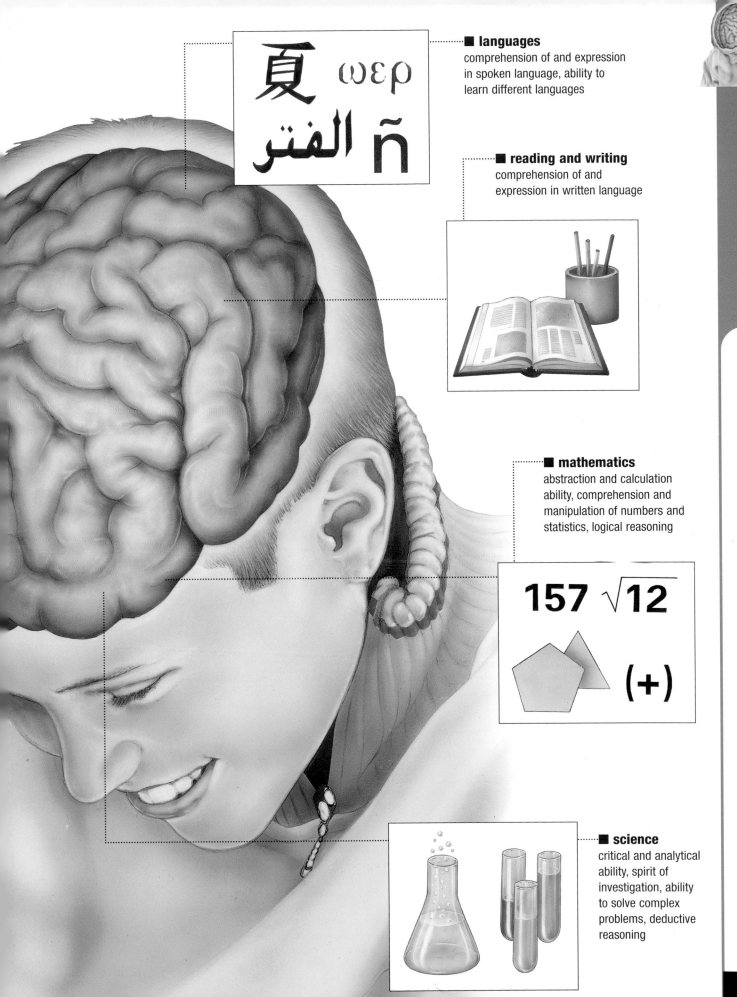

■ **languages**
comprehension of and expression in spoken language, ability to learn different languages

夏 ωελ
m̄ الفتر

■ **reading and writing**
comprehension of and expression in written language

■ **mathematics**
abstraction and calculation ability, comprehension and manipulation of numbers and statistics, logical reasoning

157 √12

(+)

■ **science**
critical and analytical ability, spirit of investigation, ability to solve complex problems, deductive reasoning

WHAT A PAIR OF NERVES!

Twelve pairs of symmetrical nerves directly connect the brain with different parts of the body, to transmit messages and carry information to and from the brain. These are the cranial pairs. They are very important because, just as some of them are an extension of the sensory organs, others are concerned with the automatic regulation of heart activity, respiration, and digestive functions.

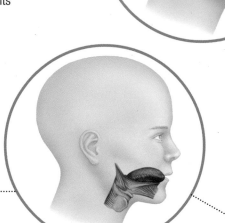

pair XII: ■
hypoglossal nerve
plays a role in the control of tongue movements

pair IX: glossopharyngeal nerve ■
conducts taste stimuli from the tongue to the brain, and takes part in the control of the pharynx muscles

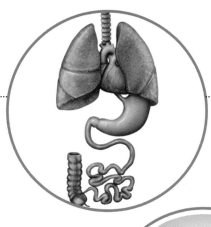

pair X: vagus nerve ■
or pneumogastric nerve
plays a role in the control of the pharynx and larynx muscles, as well as in the regulation of the heart and the respiratory and digestive organs

CENTRAL CONNECTION

Unlike the spinal nerves that exit the spinal medulla and branch out to cover the whole body, the cranial pairs rise directly from the brain, whether from the cerebrum or the encephalic truck. This difference may appear subtle, but it really is not. Cranial pairs arise in the brain because their role is of the highest priority. This is because the signals they send deal with very important actions, such as getting relevant sensory information and regulating vital functions such as heartbeat and breathing.

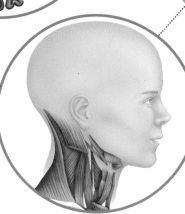

pair XI: ■
spinal nerve
plays a role in the control of the neck, shoulder, and larynx muscles

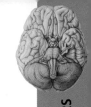

■ **pair I:**
olfactory nerve
conducts olfactory (smell-
related) stimuli from the
nose to the brain

■ **pair II: optic nerve**
conducts visual stimuli from
the eye to the brain

■ **pair III:**
oculomotor nerve
plays a role in the control
of eye movements

■ **pair IV:**
trocular nerve
plays a role in the
control of eye
movements

■ **pair VI:**
abducens nerve
plays a role in the control of
eye movements

■ **pair V:**
trigeminal nerve
conducts sensory
stimuli from the face
to the brain, and helps
control chewing

■ **pair VII: facial nerve**
conducts taste stimuli from the
tongue to the brain, and helps control
facial muscles

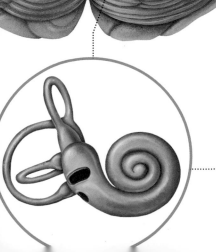

■ **pair VIII: acoustic or auditory nerve**
conducts auditory stimuli and information related
to balance from the ear to the brain

OUR BODY'S AUTOMATIC REGULATOR

The autonomic nervous system regulates the function of our bodies in an automatic and unconscious manner. It controls glandular activity, blood circulation, and many other bodily functions. This system is divided into two different and complementary sectors with opposite functions: the sympathetic nervous system, which is active when we are alert; and the parasympathetic nervous system, which dominates when we are at rest.

PARASYMPATHETIC NERVOUS SYSTEM
Controls bodily functions in relaxed, calm situations

STRESS

Stress is defined as a state of psychological and emotional tension caused by situations that cause us to be on alert, worried, or fearful because they represent a danger or threat. The result of this state may be positive, because it brings energy that is useful in confronting the problems that cause it. Stress brings about the response of the sympathetic nervous system, which prepares us for action in specific situations that demand a response. When stress is very intense or long-lasting, however, it can have a negative effect on the body, leading to exhaustion and damage to the body's systems.

eye ■
contracts the pupils

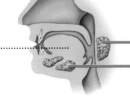

salivary glands ■
stimulates saliva production

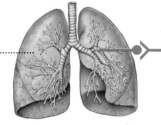

lungs ■
contracts the bronchia

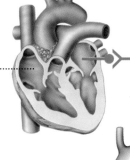

heart ■
reduces the heart rate

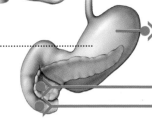

stomach ■
stimulates the secretion of gastric juices

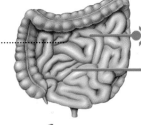

intestines ■
stimulates the digestive process

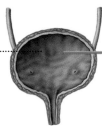

urinary bladder ■
contracts the bladder muscles

rectum ■
relaxes the muscles of the rectum

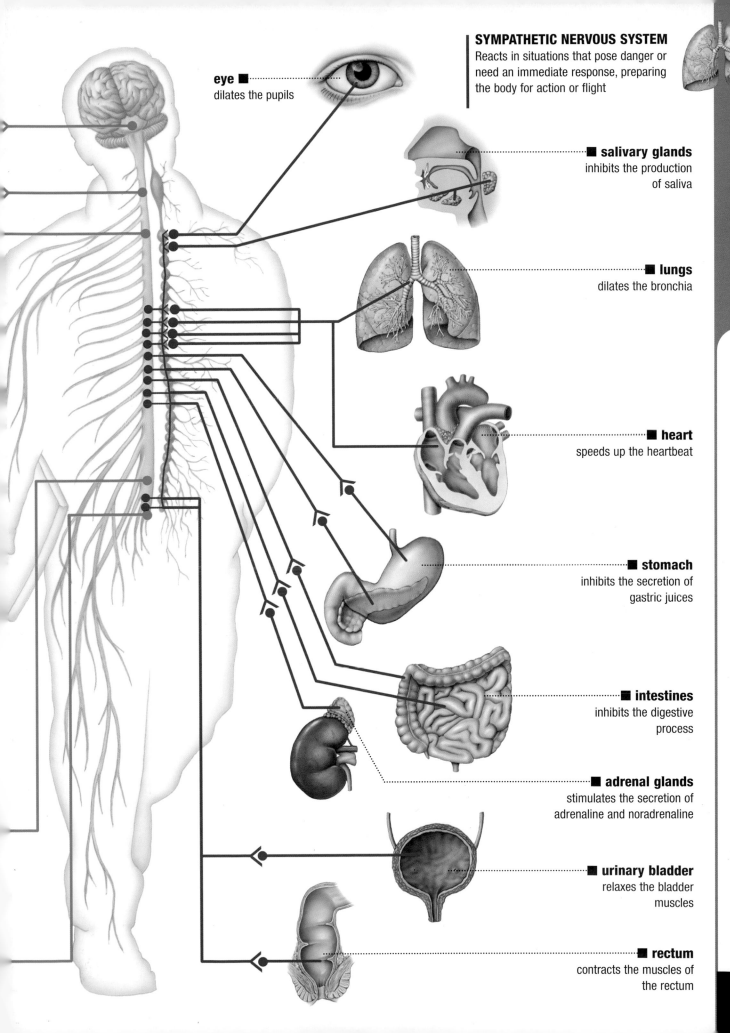

eye ■
dilates the pupils

SYMPATHETIC NERVOUS SYSTEM
Reacts in situations that pose danger or
need an immediate response, preparing
the body for action or flight

■ **salivary glands**
inhibits the production
of saliva

■ **lungs**
dilates the bronchia

■ **heart**
speeds up the heartbeat

■ **stomach**
inhibits the secretion of
gastric juices

■ **intestines**
inhibits the digestive
process

■ **adrenal glands**
stimulates the secretion of
adrenaline and noradrenaline

■ **urinary bladder**
relaxes the bladder
muscles

■ **rectum**
contracts the muscles of
the rectum

REFLEX ACTIONS

Some actions are produced automatically, without our intention, and almost without our noticing, in response to certain stimuli. These are called reflex actions. In the simplest of these, like those that occur when we suffer harm, such as when we are pricked by a pin or burned, the organs of the brain do not even get involved. The information is relayed directly to the spinal medulla, the orders for the appropriate muscle contractions are generated, and we move away from the source of harm.

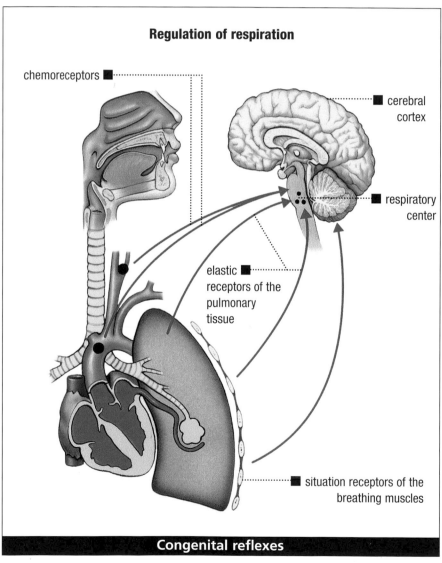

Regulation of respiration

chemoreceptors ■

■ cerebral cortex

■ respiratory center

elastic ■
receptors of the
pulmonary
tissue

■ situation receptors of the
breathing muscles

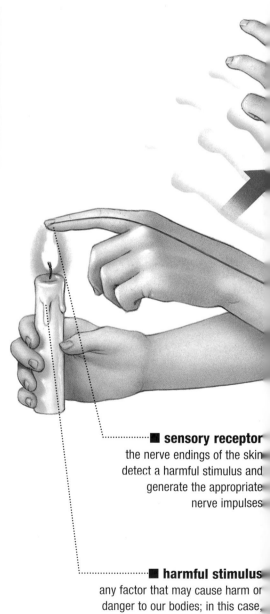

■ **sensory receptor**
the nerve endings of the skin
detect a harmful stimulus and
generate the appropriate
nerve impulses

■ **harmful stimulus**
any factor that may cause harm or
danger to our bodies; in this case,
contact with fire

Congenital reflexes

Some very important reflexes, such as those that control certain basic bodily functions like breathing or digestion, are present from birth. These complex reflexes require the participation of various brain structures, such as nerve centers at the base of the cerebrum and the encephalic trunk, but they do not need the cerebral cortex to get involved. Thus, they occur automatically, without our being aware of them. For example, we don't have to think about whether we should breathe.

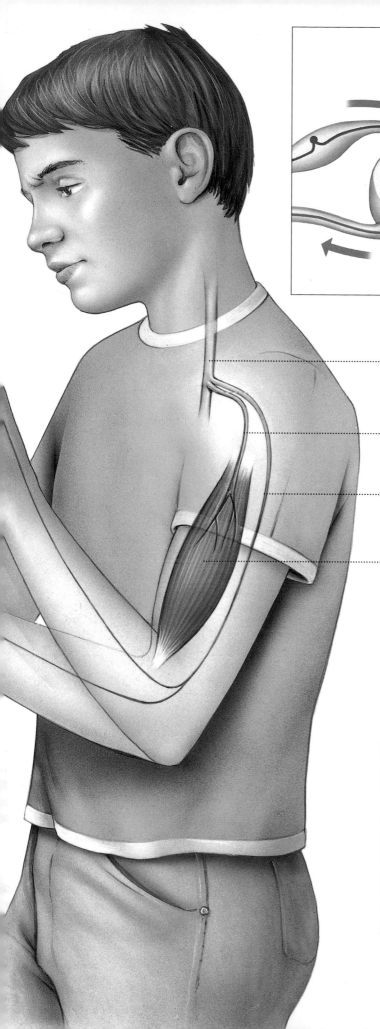

■ **sensory neuron**
conducts nerve impulses from the receptor of a harmful stimulus to the spinal medulla

■ **interneuron**
connects sensory neurons with motor neurons inside the spinal medulla

■ **motor neuron**
generates and sends appropriate orders to the target organ responsible for accomplishing them

■ **spinal cord**
simple reflex actions do not require the participation of the higher nerve structures

■ **motor nerve**
carries motor orders to the target organ

■ **sensory nerve**
carries sensory stimuli toward the spinal medulla

■ **target organ**
structure that carries out a reflex response; in this case, an arm muscle that contracts to move the hand away from the fire

CONDITIONED REFLEXES

Some reflexes are not present from birth, but are acquired over the course of life as a result of new neural pathways that form through experience. If a satisfactory response is produced by a certain stimulus, that response will occur automatically each time we encounter the stimulus that caused the reflex. For example, our mouth waters merely upon seeing our favorite dessert, because the secretion of saliva in our mouths is increased by a reflex action.

DID YOU KNOW?

INTELLIGENCE TRAITS

Experts consider intelligence the sum of a series of mental abilities that are related but independent. There are people who excel in one of these traits and others who are outstanding in a different way, making it very difficult to define and measure intelligence precisely.

Reasoning	The ability to reason and draw conclusions from known data
Verbal comprehension	Understanding the meaning of words
Verbal fluency	Knowledge of vocabulary and expressive ability
Numerical ability	Comprehension of numbers and ability to calculate
Spatial comprehension	Ability to locate oneself in space and interpret drawings and graphs
Memory	Ability to remember words, numbers, etc.
Attention span and perception	Ability to concentrate on something without getting distracted
Problem-solving ability	Ability to analyze a problem, come up with a solution, and find the means to carry it out

CURIOSITY, MOTOR OF LEARNING

Curiosity can be defined as the desire to know or to explore the unknown. It is one of the basic human characteristics that distinguishes people from many animals. Although scientists still cannot explain the brain mechanism that generates it, it is believed that curiosity depends on genetic factors, and thus has a biological basis. Psychologists think that curiosity, which is present from the first years of life, pushes us to try to learn about how the world around us works. Curiosity is a key factor in the learning process.

TYPES OF MEMORY

Memory is a marvelous ability of the human brain, since it allows us to register all kinds of information and sensations for recall a few minutes or many years later. There are three categories of memory based on how long a piece of information is retained: immediate memory, which lasts only a few seconds; short-term memory, which lasts a few hours or days; and long-term memory, which can last a lifetime.

INTERESTING FACTS

Neurons, durable cells

Neurons are the only cells in the body that do not reproduce. Those that are lost over the course of life are gone forever. However, the functions of neurons that die or are destroyed are usually taken over by other neurons that are undamaged, so that the gradual disappearance of neurons does not necessarily diminish our mental abilities, at least not until old age.

Electricity and chemistry

Nerve stimuli are sent as electrical impulses by neurons and travel through specific chemical pathways. Thus, everything we think and feel is the result of electrical flows and chemical reactions.

Electro-encephalogram (EEG)

Although the electrical impulses that travel through the nervous system are very low in voltage, they can be registered on the surface of the skin with very sensitive detectors. An electroencephalogram works by placing electrodes on the surface of the cranium to register cerebral activity.

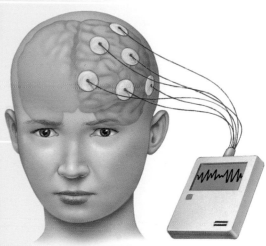

Pain, a real ally

There has been a great deal of speculation on the role of pain. Scientists believe that, contrary to what one might think considering the suffering pain causes, it can have beneficial results. Pain is like an alarm signal, a warning that something is wrong.

The evolution of the brain

Our brain is much larger than that of our closest "relative," the chimpanzee. Indeed, the size of the brain has gradually increased over the course of human evolution. While the brain of *Australopithecus* measured 17 ounces (482 g), and that of the primitive *Homo erectus* measured 34 ounces (964 g), the brain of *Homo sapiens*, modern humans, has reached an average of 50 ounces (1,418 g).

The brain of a genius

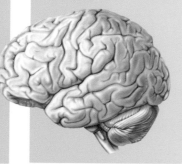

The brain of the famous physicist and mathematician Albert Einstein, creator of the theory of relativity, was turned over to science for study. Investigations showed that it had some characteristics that may have influenced Einstein's great capacity for spatial reasoning and mathematics. Although it was proven that Einstein's brain was very similar to that of most people, it was observed, among other peculiarities, that the areas related to calculation exhibited up to 15% greater development than usual.

Sleeping

The function of sleep is not precisely understood, but it seems to play a role in restoring the brain, and serves to analyze information collected during the hours we are awake. We know that sleep is a basic human need, because going without it causes a progressive deterioration of attention span and concentration, and can cause serious mental disorders when it continues for a long time. Sleep is so important that we spend no less than one-third of our lives asleep!

INDEX

The brain

J 612.82 CASSA   31057011202655

‖‖‖‖‖‖‖‖‖‖‖‖‖‖‖‖‖‖‖‖‖‖‖‖‖‖‖‖‖‖‖

Cassan, A.

WEST GA REGIONAL LIBRARY SYS